Multistep Math Problems with Written

Grade 5

Table of Contents

Multistep Math Problems with Written Explanations
Grade 5

Introduction

This book is intended to give students the opportunity to practice solving multistep computations and word problems and to explain in writing the steps that they took in the process. These skills will prepare students for standardized tests and help them to achieve the standards set forth by the NCTM (National Council of Teachers of Mathematics). These exercises assume some prior experience or familiarity with the skills involved and are designed as enrichment tools for existing curriculum.

Written explanations compel students to reconsider their work and to understand the steps they took to find their answers. In the writing, they clarify to themselves and illustrate to others that they grasp the concepts in the exercises. Due to the subjective nature of written explanations, these may vary, but the end results should be the same for all students. Students will benefit from sharing their mathematical thinking with a partner, a small group, or the entire class.

Organization
Each lesson begins with guided practice and then provides problems for skill reinforcement. The lessons can be used as classwork or homework; however, working through the practice exercises as a group will allow students to ask questions. This will help students to grasp the concepts and gain confidence before they begin working individually or in pairs.

Unit 1: Number and Operations
In this unit, students write number sentences, select appropriate operations, organize steps in a process, interpret the remainder, and choose strategies to solve problems.

Unit 2: Algebra
In this unit, students make a model, choose a method, and write number sentences using variables to represent unknown quantities.

Unit 3: Geometry
Students solve multistep problems dealing with perimeter, circumference and area of circles, and areas of squares, rectangles, and triangles by making a sketch and using formulas and estimation.

Unit 4: Measurement
In this unit, students solve problems concerning capacity, weight, making a schedule, and planning a budget. Problems ask students to choose a variable or work problems that require many steps.

Unit 5: Data Analysis and Probability
Students study line, bar, and circle graphs, as well as explore combinations and sampling.

Assessments
There are two kinds of assessments.
- There is a general assessment that covers important material appropriate for the fifth grade on pages 3 and 4. It can be given as a pretest to gauge students' knowledge of the material to be covered in this book. Later in the year, the same test can be administered to determine students' understanding, progress, and achievement.
- Each unit also has an assessment. These unit assessments can be administered at any time during the unit as a pretest, review, or posttest for specific concepts.

Special Note
Students may need to use their own paper to work some of the problems and explain the steps they took to arrive at their answers.

Correlation to NCTM Standards
The NCTM has set specific standards to help students become confident in their mathematical abilities. Multistep problems and written explanations are important components of the mathematics curriculum because they represent an important extension of students' knowledge and understanding of mathematics. A thorough understanding of patterns, equations, representations, and variables heightens students' awareness of the usefulness of mathematics in everyday life. This book is designed to help parents and teachers guide students toward achievement of the standards for their grade level through problem solving, reasoning and proof, communication, connections, and representation. According to NCTM standards, students should be able to:
- build new mathematical knowledge through problem solving;
- solve problems that arise in mathematics and in other contexts;
- apply and adapt a variety of appropriate strategies to solve problems;
- organize and consolidate their mathematical thinking through communication;
- communicate their mathematical thinking coherently and clearly to peers, teachers, and others;
- analyze and evaluate the mathematical thinking and strategies of others;
- use the language of mathematics to express mathematical ideas precisely.

Name _____ Date _____

Overall Assessment

Directions Solve each problem. Explain the steps you took. ✐

1. Sally wants to buy 3 shirts that cost $12.99 each, 2 pairs of jeans that cost $24.99 each, and a hat that costs $9.99. She has $150.00. Does she have enough money?

   _____

   _____

   _____

   _____

2. Kristen is selling cookies. She has sold 25 boxes. She needs to sell a total of 86 boxes to win a prize. Write an equation with a variable to find out how many more boxes she needs to sell.

   _____

   _____

   _____

   _____

3. Kevin wants to buy a school lunch for $1.75. He has 5 quarters, 3 dimes, 4 nickels, and 5 pennies. Does he have enough money? Make a model to solve the problem.

   _____

   _____

   _____

   _____

4. Jessica has 11 coins in her pocket worth $0.94. She has some quarters, dimes, nickels, and pennies. She has the same number of quarters as nickels. She has one more dime than nickels and one more penny than dimes. How many of each does she have?

   _____ quarters _____ dimes _____ nickels _____ pennies

Go on to the next page.

www.svschoolsupply.com
© Steck-Vaughn Company
3
Overall Assessment
Multistep Math Problems with Written Explanations, Grade 5, SV 6128-X

Overall Assessment, page 2

Directions Solve each problem. Explain the steps you took. ✏️

5. Jeff has a circular pond in his backyard. He wants to build a square garden around it. If the pond has a diameter of 6 yards and the garden is to be 9 yards on each side, how many square yards will the garden be?

_____

_____

_____

6. **a.** Karen is wallpapering her bathroom. Two walls are 2.25 meters wide and 2 other walls are 3.25 meters wide. All 4 walls are 3 meters high. Find the total area of all the walls in her bathroom.

_____

_____

_____

b. If 1 roll of wallpaper is 0.75 meters wide and 8 meters long, how many rolls of wallpaper should Karen buy?

_____

_____

_____

7. Each week, Ethan's dog eats 112 ounces of dog food. How many pounds per day is this?

_____

_____

_____

Unit 1 Assessment

Directions Solve each problem. Explain the steps you took. ✎

1. Erica wants to buy some plants for her backyard. She picks out 2 rosebushes for $12.99 each, 3 pallets of grass for $3.99 each, and an oak tree for $39.99. How much money will she spend?

_____

_____

_____

2. Chris is packing up his old books to give to the neighborhood garage sale. The boxes he is using hold 5 books each. If he has 38 books to give away, how many boxes will he use?

_____

_____

_____

3. Stephanie ran a total of 35 miles in 3 days. She ran 11 miles on Monday and 12 miles on Wednesday. How far did she run on Friday?

_____

_____

_____

4. Aaron is borrowing $50 from Seth to buy new skateboard equipment. Seth says Aaron can either pay $3 per week for 20 weeks or $10 per week for 5 weeks. Which plan should Aaron choose? Why?

_____

_____

_____

Add, Subtract, Multiply, or Divide

Often, you must add, subtract, multiply, or divide to solve a problem. Look for what the problem asks you to do. If it asks you to put together or join groups, then you usually add or multiply. If it asks you to take away or to find out how many in each group, then you usually subtract or divide. Read the problem carefully because you may have to do more than one operation to find the answer.

Here's How!

Step 1: **Read the problem.**
Luisa, Melvin, and Guillermo are potting plants to sell at the sixth-grade plant sale. Luisa has potted 20 zinnias, Melvin has potted 13 petunias, and Guillermo has potted 16 marigolds. Their group needs to pot 60 plants in all. How many plants still need to be potted?

Step 2: **Look for what the problem asks you to do.**
It asks you to find out how many plants still need to be put in pots. First, add to find out how many plants have already been put into pots. Next, subtract to find out how many still need to be potted.

Step 3: **Solve the problem.**
$20 + 13 + 16 = 49$ $60 - 49 = 11$
They still need to put 11 plants in pots.

Directions Solve each problem. Explain the steps you took. ✏

1. The class had already potted 25 marigold plants, 30 zinnia plants, and 20 petunia plants. However, a shelf full of plants just fell down, and 8 zinnias need to be repotted. How many plants are ready for the sale now?

_____

2. The class wants to earn $600 from the plant sale. If the class sells 55 marigolds at $4 each and 55 zinnias at $5 each, how much do the students need to charge for the 35 petunias they have for sale?

_____

3. The sale is in full swing, and the students have already sold 19 of the 55 marigolds, 16 of the 35 petunias, and 28 of the 55 zinnias. How many plants do the students have left to sell altogether? _____

More Than One Operation

Sometimes, in order to solve a problem, you have to use more than one operation. Read the problem carefully. Decide what the problem is asking you to do. Choose the steps you will take. Then, solve the problem.

The fifth-grade class is hosting a costume party. There will be prizes for the most original costume, the funniest costume, and the scariest costume.

Here's How!

Step 1: Read the problem.

For her costume, Maria decides to dress up as a potted fern. She buys a roll of foam rubber 12 meters long and 1 meter wide. It takes 2 meters of foam rubber to make 1 leaf. How many leaves can she make and still have 2 meters left for the base of her costume?

Step 2: Identify the facts.

Maria needs to save 2 meters of foam for the base of her costume. She has 12 meters of foam. It takes 2 meters to make 1 leaf.

Step 3: Solve the problem.

Subtract 2 meters from the total amount of foam.

$12 - 2 = 10$ meters

Divide the remaining foam by 2 meters to determine how many leaves she will be able to make.

$10 \div 2 = 5$

She will be able to make 5 leaves.

Directions Solve each problem. Explain your answer. ✏️

1. James, Elizabeth, Karen, Richard, Steve, and Marvin come to the party together. Richard leaves early, but 2 friends of Karen's join the group. By the end of the party the group is half its original size. What is the total number of people who left the group?

_____

_____

Go on to the next page.

More Than One Operation, page 2

2. Mark decides to dress as his favorite food. He makes meatballs out of papier mâché and noodles out of ribbon. The ribbon comes in spools 8 meters long. If Mark makes each noodle 2 meters long, how many spools of ribbon will he need to make 45 noodles?

_____

_____

3. Christina comes to the party dressed as a caterpillar, and she shares the prize for the funniest costume with Mark. Since no one dressed in a scary costume, the judges decide to combine the $10 prize for scariest costume with the $15 prize for funniest costume, and to divide it equally between Christina and Mark. How much money will each one receive?

_____

_____

4. Alfie and Marie are tied for the $25 prize for the most original costume. They give some of the money to Bobby, who helped them make the costumes, and divide the rest evenly. If they each keep $8.50, how much money did they give Bobby?

_____

_____

Name _____ Date _____

Solving Two-Step Problems

Two families take a long-distance hike from the town of Redbird to Sprucewood State Park. Each family takes a different route. On normal ground, the hikers cover 10 miles per day. In the mountains, they travel about 7 miles per day. The map shows the distance and terrain of each route.

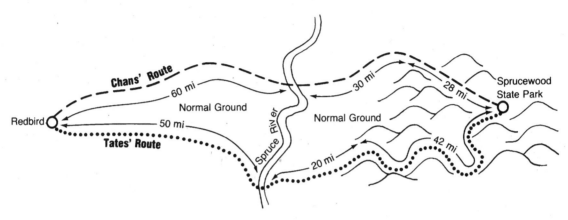

Directions Write in the missing step. Then, solve.

1. The Chans walk over normal ground for 4 days. How many more miles must they walk to reach the river?

   Step 1: _____

   Step 2: Subtract to find how many miles are left before they reach the river.

   _____

   Answer: _____

2. On the Tates' route, how many more miles is it from the river to the park than it is from Redbird to the river?

   Step 1: Add to find how many miles it is from the river to the park.

   _____

   Step 2: _____

   Answer: _____

Go on to the next page.

Solving Two-Step Problems, page 2

Two families take a long-distance hike from the town of Redbird to Sprucewood State Park. Each family takes a different route. On normal ground, the hikers cover 10 miles per day. In the mountains, they travel about 7 miles per day. The map shows the distance and terrain of each route.

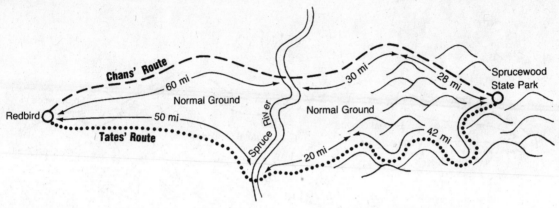

3. A shortcut could get the Chans to the river in just 4 days. Walking at their usual rate, how many miles would they save?

_____

_____

4. One family has walked through mountains for 4 days. They are now 14 miles from the park. Which family is it?

_____

_____

5. The Tates have traveled from Redbird for 7 days. Have they reached the mountains yet?

_____

_____

6. According to the map, how many miles longer is the Chans' route than the Tates' route?

_____

_____

Name _____ Date _____

Step by Step

When a problem requires more than one step to solve, organize the facts you know. Then, take one step at a time.

Here's How!

Step 1: Read the problem.

Compustore has 75 computers in stock. On Monday they sell 8. On Tuesday they sell 6. How many computers do they have left to sell?

Step 2: Organize the facts.

The store has 75 computers.
They sell 8, then they sell 6.

Step 3: Solve the problem.

Add to find how many they sell altogether. $8 + 6 = 14$
Subtract to find how many they have left. $75 - 14 = 61$
They have 61 computers left to sell.

Directions Solve each problem. Explain the steps you took.

1. Amando saves $1,000 to buy a computer. He buys one for $895. The sales tax is $53.70. How much money does Amando have left?

   _____

   _____

   _____

2. Anna has $50.00 to spend at Compustore. She buys a software program that costs $32.50. On her way out, she remembers that she needs a new mouse. She pays $9.75 for her new mouse. How much does she have left?

   _____

   _____

   _____

Go on to the next page.

Name _____ Date _____

Step by Step, page 2

When a problem requires more than one step to solve, organize the facts you know. Then, take one step at a time.

Directions Solve each problem. Explain the steps you took. ✏️

3. Jeanine buys a box of computer paper for $9.95, a magazine for $1.95, and disks for $6.99. She pays with a $20 bill. How much change does she get?

   _____

   _____

   _____

4. Elena and Ethan share a box of 100 computer disks. She has used 21 disks, and he has used 27 disks. How many computer disks do they have left?

   _____

   _____

   _____

5. Angela wants to buy a printer that costs $175.95. She has saved $85.37. She had earned $48.25 working part-time. How much more money does she need?

   _____

   _____

   _____

6. Sara wants to buy 3 computer games. Two of the games cost $19.95 each and one costs $29.95. If she has $100, will she have enough money?

   _____

   _____

   _____

Interpreting the Remainder

Imagine that you are an astronaut on a mission in space. You must find out if the planet Dither can support a space colony. You must take samples of the soil and air, photograph the landscape, and conduct experiments.

Here's How!

Step 1: **Read the problem.**

The soil tester deposits 9 soil samples at a time in lab trays. There are 321 samples to be tested. How many lab trays will be used? Will all these lab trays be completely full?

Step 2: **Make a plan.**

Divide to find out how many trays will be used.
$321 \div 9 = 35 \text{ R } 6$

Step 3: **Solve. Interpret your answer.**

Since there is a remainder, one tray will not be completely full. This answer means that there will be 35 full trays and one tray with 6 samples. 36 trays will be used.

Directions ➤ Solve each problem. Explain what the remainder tells you. ✏️

1. There are 42 air tanks to be taken to a testing site. If 5 air tanks fit into a tank carrier, how many full carriers will there be? How many tanks will be left over and then put into a partly full carrier?

   _____

   _____

   _____

2. Each air tank has 16 hours of air in it. If each experiment takes 3 hours, how many experiments can you complete before switching tanks? Will any air be left over?

   _____

   _____

   _____

Go on to the next page.

Interpreting the Remainder, page 2

Imagine that you are an astronaut on a mission in space. You must find out if the planet Dither can support a space colony. You must take samples of the soil and air, photograph the landscape, and conduct experiments.

Directions ➤ Solve each problem. Explain what the remainder tells you. ✏

3. Your space car can carry 8 pounds of rock samples on each trip. If you have 68 pounds of samples to take back to the ship, how many trips must you make in the space car?

   _____

   _____

   _____

4. It takes 4 fuel packs to get the space car to and from Testing Site X. If you have 35 fuel packs, how many times can you go to and from the testing site? How many fuel packs will be left over?

   _____

   _____

5. The landscape camera holds enough film to take 60 pictures. One day, Mission Control sends you out to take pictures in groups of 7. How many groups of 7 pictures will you be able to take? How many pictures could you take with the film that is left over?

   _____

   _____

6. You and the 3 other members of your crew must conduct 22 experiments in the 7 days you are going to be on the planet. You can conduct the same number of experiments each day. How many experiments is that per person? How many per day? If one of the leftover experiments was given to each person, how many people would have to do an extra experiment? How many experiments would be done on the last day?

   _____

   _____

Choosing a Strategy to Solve Multistep Problems

Different strategies work best to solve different problems. You may want to guess and check, draw a picture, or work backwards. Or you may find that there is extra information in the problem. Decide what the problem is asking, then choose a strategy to solve.

Here's How!

Step 1: Read the problem.

Marla's paper was due in 5 days. In order to finish it, she worked 4 hours on Sunday, 1 hour on Monday, and 2 hours on Tuesday. She finished it on Wednesday. If she spent a total of 10 hours on her paper, how many hours did she work on Wednesday?

Step 2: Decide what the problem is asking. Make a plan.

In order to find out how many hours she worked on Wednesday, add the number of hours she spent working each day and subtract from the total. You do not need the fact that Marla's paper was due in 5 days. It does not give any information about how many hours she worked.

Step 3: Solve.

4 hours + 1 hour + 2 hours = 7 hours

10 hours − 7 hours = 3 hours

Marla worked 3 hours on Wednesday.

Directions Choose a strategy to solve each problem. Explain the steps you took. ✐

1. Daniel and Angelo worked together for 3 hours on a school project. Angelo spent $4.60 on supplies. Daniel spent $3.90. How much money does Daniel owe Angelo so they will spend the same amount?

   _____

2. In the first race, the racing cars had to go around a 250-meter track 7 times. In the second race, they had to go around the track 5 times. How many meters longer was the first race than the second?

   _____

Name _____ Date _____

Unit 2 Assessment

Directions Solve each problem. Explain your answers. ✐

1. Bo and Juan are saving money to go to camp. Bo has saved 3 $5-bills,
   6 $1-bills, 3 quarters, and a dime. Juan has saved 4 $5-bills, 3 $1-bills,
   5 quarters and 3 pennies. How much money do they have altogether?

   _____

   _____

   _____

2. Michelle has 12 coins. She has some pennies, nickels, and dimes. She has
   6 times as many nickels as pennies, and 1 less dime than nickels. She has only
   1 penny. What coins does she have?

   _____

   _____

   _____

Directions Choose a variable, write a number sentence, and solve. ✐

3. Sam has 4 fewer baseball caps than Eric has. If Sam has 12 caps, how many
   does Eric have?

   Let _____ = _____

   Number sentence _____

   Answer: _____

4. Colin earns $4 each day for walking the neighborhood dogs. If he earned $64
   last month, how many days did he walk the dogs?

   Let _____ = _____

   Number sentence _____

   Answer: _____

Name _____ Date _____

Making a Model

Sometimes it is helpful to make a model or diagram to solve problems.

Here's How!

Step 1: Read the problem.

Ben and Soo Lin want to buy a small pizza to share. The pizza costs $4.99. Ben has 2 $1-bills, 1 quarter, and 1 dime. Soo Lin has 2 $1-bills, 2 quarters, 1 nickel, and 3 pennies. Do they have enough money to buy the pizza?

Step 2: Make a diagram.

Ben's money   

Soo Lin's money 

Step 3: Find out how much money each person has.

Ben: $2.35   Soo Lin: $2.58

Step 4: Add their money together.

$2.35 + $2.58 = $4.93

Step 5: Answer the question.

No. They have $4.93. That is not enough for the pizza.

Directions Make a model to answer each question.

1. Stephen and Brett are running errands to make some extra money. Stephen has 2 $5-bills, 4 $1-bills, 4 quarters, 6 dimes, and 2 nickels. Brett has 3 $5-bills, and 8 dimes. Who has earned more?

2. Dave and Pedro are saving to buy a model ship. Dave has saved 2 $1-bills, 3 quarters, 5 dimes, and 2 pennies. Pedro has saved 1 $1-bill, 5 quarters, 1 nickel, and 5 pennies. How much have they saved altogether?

Go on to the next page.

Making a Model, page 2

Sometimes it is helpful to make a model or diagram to solve problems.

Directions Make a model to answer each question. ✐

3. Maya is at the bowling alley. It costs $1.75 to rent shoes and $3.25 to bowl 1 game. She has 3 $1-bills, 6 quarters, 4 dimes, 2 nickels, and 3 pennies. Does she have enough money to go bowling?

4. Soymilk costs $1.99 for a half-gallon. Jeff is at the register with 2 half-gallons when he realizes that he left his wallet at home. He empties his pockets and comes up with 1 $1-bill, 7 quarters, 1 dime, 1 nickel, and 6 pennies. How many half-gallons of soymilk can he buy, if any?

5. Nabil paid $2.35 for a school lunch. He gave the cashier 3 $1-bills. Later he went back to the counter to buy dessert. The cookies cost $0.30 each, and the ice cream sandwiches cost $0.85 each. Was Nabil able to buy any dessert with the change he had received? What could he buy?

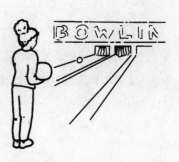

Choosing Your Method

> There are many ways to solve difficult problems. You can make a drawing, make a table, use guess-and-check, or make up your own method.

Directions Solve each problem. Explain your method. ✏️

1. Kazuko has $1.30 in nickels, dimes, and quarters. She has twice as many nickels as dimes and twice as many dimes as quarters. How many of each coin does she have?

_____ nickels   _____ dimes   _____ quarters

_____

_____

_____

2. Jeb has $5.00 worth of coins, but none of them are nickels or quarters. He has 1 half-dollar. He has a total of 100 coins. How many dimes and pennies does he have?

_____ dimes   _____ pennies

_____

_____

_____

3. Nikki has 6 coins. She has some dimes, nickels, and pennies. She has at least 1 of each type of coin. She has more nickels than dimes and more dimes than pennies. What is the total value of her coins?

value: _____

_____

_____

_____

Go on to the next page.

Choosing Your Method, page 2

There are many ways to solve difficult problems. You can make a drawing, make a table, use guess-and-check, or make up your own method.

Directions Solve each problem. Explain your method.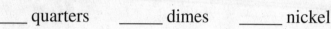

4. Ruth has twice as many dimes as nickels, and 5 times as many nickels as quarters. She has $3.00 in coins in all. How many of each coin does she have?

_____ quarters   _____ dimes   _____ nickels

_____

_____

_____

5. Reggie has $2.00 worth of nickels, dimes, and quarters. He has a total of 20 coins. He has the same number of dimes as quarters. How many of each coin does he have?

_____ quarters   _____ dimes   _____ nickels

_____

_____

_____

6. Brad has $2.50 worth of 3 kinds of coins. The largest coins are worth less than the smallest ones, but their total value is twice as much as the total value of the smallest ones. The medium-sized coins have a total value of 2 of the largest coins. What are the coins and how many of each are there?

_____

_____

_____

Name _____ Date _____

Writing Number Sentences

Sometimes, writing a number sentence to represent a situation can help you solve a problem.

Here's How!

Step 1: Read the problem.

Theo has 23 cousins. If 16 of his cousins are girls, how many of his cousins are boys?

Step 2: Choose a variable to represent what you need to find.

Let n equal the number of cousins that are boys.

Step 3: Write a number sentence.

boy cousins + girl cousins = all cousins
$$n \quad + \quad 16 \quad = \quad 23$$

Step 4: Solve the problem.

Subtract the same number from both sides of the number sentence.

$$n + 16 = 23$$
$$n + 16 - 16 = 23 - 16$$
$$n = 7$$

So, Theo has 7 cousins that are boys.

Directions Choose a variable, write a number sentence, then solve. Be sure to explain what the variable stands for.

1. Alayna has 30 beads. She has 8 blue beads, and the rest are red. How many red beads does she have?

   Let _____ = _____

   Number sentence _____

   Answer: _____

Go on to the next page.

Name _____ Date _____

Writing Number Sentences, page 2

Directions Choose a variable, write a number sentence, then solve. Be sure to explain what the variable stands for. ✏️

2. Chelsea weighs 9 pounds more than Tom. If Chelsea weighs 87 pounds, how much does Tom weigh?

   Let _____ = _____

   Number sentence _____

   Answer: _____

3. Julia worked 5 more hours in the school library during October than in September. She worked 15 hours in October. How many hours did she work in September?

   Let _____ = _____

   Number sentence _____

   Answer: _____

4. Jason likes to read every evening. He read 3 chapters on Friday. By Sunday morning he had read 8 chapters. How many chapters did he read on Saturday?

   Let _____ = _____

   Number sentence _____

   Answer: _____

5. Stephen received only 86 points for his book report. He lost 4 points for turning it in a day late. How many points would he have received if his report had been on time?

   Let _____ = _____

   Number sentence _____

   Answer: _____

Name _____ Date _____

Number Sentences

> A number sentence can help you solve problems.

When the Sanders bought a new motorboat, they started keeping a ship's log. One day while they were in heavy surf, water splashed on the pages, and the ink ran. Answer the questions to complete the ship's log.

Here's How!

Step 1: Read the problem.

We took our new boat on our first overnight fishing trip. We motored 45 miles Saturday. We motored _____ miles Sunday, and 102 miles in all.

Step 2: Choose a variable to represent what you need to find.

Let m equal the number of miles motored Sunday.

Step 3: Write a number sentence.

$45 + m = 102$

Step 4: Solve and answer the question.

$m = 102 - 45$

$m = 57$

The Sanders motored 57 miles on Sunday.

Directions ➤ **Choose a variable, write a number sentence, and solve each problem. Be sure to explain what the variable stands for.** ✏️

1. Scott took the boat to the marina today and spent $26 on gasoline and oil. The gasoline cost $18, and the oil cost $_____.

   Let _____ = _____

   Number sentence _____

   Answer: _____

Go on to the next page.

Number Sentences, page 2

Directions Choose a variable, write a number sentence, and solve each
problem. Be sure to explain what the variable stands for. ✏️

2. Hank likes to fish. He just bought 4 sets of deep-sea
fishing tackle for $52 each, or $_____ altogether.

Let _____ = _____

Number sentence _____

Answer: _____

3. Scott is trying to economize. Instead of buying new ski ropes, he made his own.
He bought 225 feet of rope and cut it into 3 equal pieces. Now we have new
ropes, each one _____ feet long.

Let _____ = _____

Number sentence _____

Answer: _____

4. We intend to take our time sailing to New Orleans. We have 81 miles to travel
and 9 days to cover the distance. We only need to travel _____ miles per day.

Let _____ = _____

Number sentence _____

Answer: _____

5. Today, Hank clocked the boat's speed. Our cruising speed is 11.2 miles per
hour. In 5 hours, we could travel _____ miles.

Let _____ = _____

Number sentence _____

Answer: _____

More Number Sentences

You can find an unknown number by using a variable to write a number sentence.

The Drones and UBX-42 are two new music groups whose CDs are competing for first place on the music charts. Answer the questions to see how well they are doing.

Here's How!

Step 1: Read the problem.

In the recording studio, the Drones taped a song that lasted 48.7 seconds. They were not satisfied with it, so they taped it again. The second taping lasted 2.9 seconds less than the first. The third taping lasted 0.47 seconds less than the second. How long was the third taping?

Step 2: Choose a variable.

Let x equal the length of the third taping.

Step 3: Write a number sentence.

$x = 48.7 - 2.9 - 0.47$

Step 4: Solve and answer the question.

$x = 45.33$

The third taping lasted 45.33 seconds.

Directions Write a number sentence for each problem and solve. Explain how you found each answer.

1. The Drones made a music video of their hit single. Their video was 2 minutes 23.3 seconds longer than the current UBX-42 video. The UBX-42 video lasts 6 minutes 32 seconds. How long is the Drones' video?

   Number sentence _____

   Answer: _____

   _____

   _____

   _____

Go on to the next page.

More Number Sentences, page 2

Directions Write a number sentence for each problem and solve. Explain how you found each answer. ✎

2. The Drones' CD has sold 1.35 million copies in 8 months. Their manager said he expects the CD to sell another 0.42 million in the next 2 months. By the end of the year, he thinks the CD will have sold 0.74 million more copies than in the first 10 months. How many copies might be sold altogether in 12 months?

Number sentence _____

Answer: _____

_____

_____

_____

3. On the first night of a 3-concert tour, the Drones drew 47,320 fans. At the second concert, 2,420 fewer fans attended than the first concert. At the third concert, 1,755 more fans attended than the second concert. How many fans were at the third concert?

Number sentence _____

Answer: _____

_____

_____

_____

Name _____ Date _____

Unit 3 Assessment

Directions Solve each problem. You may need to draw a picture. Explain how you found each answer.

1. Luci is working on a jigsaw puzzle of the Alamo. When it is finished, she would like to frame it. If the puzzle measures 14 in. × 16 in., how many inches of frame will she need?

   _____

   _____

2. The diameter of a circle is 10 inches. Find the circumference and area.

   _____

   _____

3. The neighborhood playground is circular with a square-shaped garden surrounding the playground. The diameter of the circular playground is 20 feet. The square garden measures 30 feet on each side. What is the area of the garden around the playground?

   _____

   _____

4. Frank is planning a garden. He wants to make the garden 20 feet long and 10 feet wide. What will the area of the garden be?

   _____

   _____

Name _____ Date _____

Finding the Perimeter

To find the perimeter, add the total distances around a figure.

Here's How!

Step 1: **Read the problem.**
The square cattle pen is 75 feet on each side. The rectangular horse pen is 250 feet by 300 feet. How much longer is the fence around the horse pen than the fence around the cattle pen?

Step 2: **Find the perimeter of the cattle pen.**
75 + 75 + 75 + 75 = 300 feet

Step 3: **Find the perimeter of the horse pen.**
250 + 250 + 300 + 300 = 1,100 feet

Step 4: **Answer the question.**
Subtract the perimeter of the cattle pen from the perimeter of the horse pen.
1,100 – 300 = 800 feet
The fence around the horse pen is 800 feet longer.

Directions Answer the questions. Explain the steps you took.

1. The playground is 110 feet long on the north side, 232 feet on the south side, 155 feet on the east side, and 280 feet long on the west side. Anne walked around the perimeter of the playground 4 times. How far did she walk?

   _____

2. Jim wants to build a rectangular sandbox. He has 4 boards. They are 5 feet long, 6 feet long, 8 feet long, and 9 feet long. What is the largest size he can make the sandbox? What will the perimeter of the sandbox be?

   _____

3. Sara wants to fence in a triangular section of land that is 45 feet on each side. If fencing is sold in 5-foot sections, how many sections will she need to buy?

   _____

Name _____ Date _____

Circumference and Area of a Circle

To find the circumference of a circle, use the formula C = π × d, where d is the diameter and π is equal to 3.14. The radius is $\frac{1}{2}$ the diameter. To find the area of a circle, use the formula A = π × r^2 , or A = π × r × r, where r is the radius.

Here's How!

Step 1: **Read the problem.**
Find the circumference and area of a circle with a diameter of 8 inches.

Step 2: **Identify the facts.**
The circle's diameter is 8 inches, so the radius is 4 inches.

Step 3. **Use the formulas.**

C = π × D   A = π × r × r
C = 3.14 × 8 = 25.12   A = 3.14 × 4 × 4 = 50.24
The circumference is   The area is
25.12 inches.   50.24 square inches.

Directions Make a drawing and use the formulas to help you solve these problems. ✐

1. The diameter of a circle is 18 inches.
   Find the circumference and area.

   Circumference _____

   Area _____

2. A circle has a diameter of 16 inches.
   What is the circumference and area of the circle?

   Circumference _____

   Area _____

Go on to the next page.

Circumference and Area of a Circle, page 2

To find the circumference of a circle, use the formula C = π × d, where d is the diameter and π is equal to 3.14. The radius is ½ the diameter. To find the area of a circle, use the formula A = π × r², or A = π × r × r, where r is the radius.

Directions Make a drawing and use the formulas to help you solve these problems. ✎

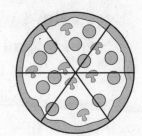

3. A cake has a radius of 6 inches.
   What is the area of the cake?

   Area _____

4. An apple pie has a radius of 5 inches.
   What is the area of the pie?

   Area _____

5. If you cut the pie into 8 pieces, how many
   square inches of pie will each piece be?

   Answer _____

6. A tabletop has a circumference of 188.4 inches.
   What is the radius of the tabletop?

   Radius _____

Circles in Squares

Some problems require that you find information using different formulas.
Helpful formulas: Area of square = side × side
Area of circle = π × radius × radius
Use π = 3.14

Here's How!

Step 1: Read the problem.
One circle with a 5-cm radius fits in this square. The area of the square is how much greater than the area of the circle?

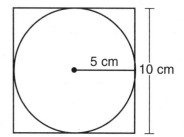

Step 2: Find the area of the square.
Area of a square 10 × 10 = 100 sq. cm

Step 3: Find the area of the circle.
Area of the circle: 3.14 × 5 × 5 = 78.5 sq. cm

Step 4: Answer the question.
Subtract the area of the circle from the area of the square.
100 – 78.5 = 21.5 sq. cm
The square is 21.5 sq. cm larger than the circle.

Directions Answer the questions. Explain the steps you took. ✎

1. One circle with a 12-cm radius fits in this square. The area of the square is how much greater than the area of the circle?

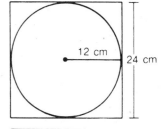

_____

2. Four circles, each with a 6-cm radius, fit in the square. The area of the square is how much greater than the combined areas of the circles?

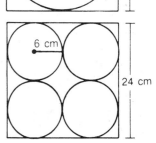

_____

Go on to the next page.

Circles in Squares, page 2

Helpful formulas: Area of square = side × side
Area of circle = π × radius × radius
Use π = 3.14

Directions Answer the questions. Explain the steps you took. ✎

3. Nine circles, each with a 4-cm radius, fit in the square. The area of the square is how much greater than the combined areas of the circles?

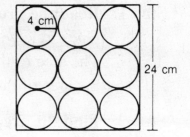

_____

_____

4. How many circles with a 3-cm radius will fit in the square?

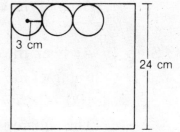

_____

_____

5. The area of the square is how much greater than the combined areas of the circles with 3-cm radii?

_____

_____

6. How many circles with a 2-cm radius will fit in the square?

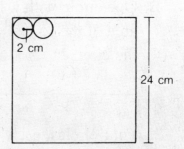

_____

_____

7. What pattern do you see after working problems 1 through 6?

_____

_____

Making a Drawing to Find Area

> Drawing a picture is a helpful way to solve problems.
>
> Helpful formulas: Area of rectangle = length × width
>
> Area of square = side × side
>
> Area of triangle = $\frac{1}{2}$ base × height
>
> Area of circle = π × radius × radius
>
> Area of semicircle = $\frac{1}{2}$ π × radius × radius
>
> Use π = 3.14
>
> Radius = $\frac{1}{2}$ diameter

Here's How!

Step 1: Read the problem.

Mrs. Smollet made a table that has 4 folding leaves. The center of the table is square, and each leaf is a semicircle whose diameter is equal to 1 side of the square. The square is 4 feet per side. What is the area of the table?

Step 2: Make a drawing.

Step 3: Make a plan.

The table has 1 square and 4 semicircles.

Find the area of each piece and add them together.

Step 4: Solve and answer the question.

Area of square: 4 × 4 = 16   Area of semicircle = $\frac{1}{2}$ × 3.14 × 2 × 2 = 6.28

Since there are 4 semicircles, multiply 6.28 × 4 = 25.12.

16 + 25.12 = 41.12. The area of the table is 41.12 square feet.

Directions Use a drawing to solve each problem. Explain the steps you took. ✐

1. Milo has a round vegetable garden enclosed by a low, square wall. The wall is 24 feet long on each side. What is the area of the vegetable garden?

_____

Go on to the next page.

Making a Drawing to Find Area, page 2

Drawing a picture is a helpful way to solve problems.
Helpful formulas: Area of rectangle = length × width
Area of square = side × side
Area of triangle = $\frac{1}{2}$ base × height
Area of circle = π × radius × radius
Use π = 3.14
Radius = $\frac{1}{2}$ diameter

Directions Use a drawing to solve each problem. Explain the steps you took. ✎

2. Lee is a designer. The sketch that he is working on is a rectangle that is 13 inches by 15 inches. One of the 13-inch sides is also the base of a triangle that has a height of 10 inches. One of the 15-inch sides is also the base of a triangle that has a height of 12 inches. What is the total area of the sketch?

_____

3. In front of an office building there is a round pond that has a flower garden around it. The diameter of the entire garden is 18 feet. The diameter of the pond is 6 feet. What is the area of the garden around the pond?

_____

4. A large sculpture in North Shore Seaport's town square is triangular in shape. Its area is 6,405 square feet. Its base is 61 feet wide. A similar sculpture in South Shore has an area of 11,505 square feet and a base that is 59 feet wide. How much taller is the sculpture in South Shore than the one in North Shore?

_____

5. The McGraths' new property is 200 feet long and 75 feet wide. Their previous property was 160 feet long and 80 feet wide. Is the area of the McGraths' new property greater or less than that of their old property? By how much?

_____

Name _____ Date _____

Using Drawings and Formulas

Drawing a picture is a helpful way to solve problems.

Helpful formulas: Area of rectangle = length × width

Area of square = side × side

Area of triangle = $\frac{1}{2}$ base × height

Area of semicircle = $\frac{1}{2}$ π × radius × radius

Use π = 3.14

Radius = $\frac{1}{2}$ diameter

Here's How!

Step 1: **Read the problem.**

Step 2: **Make a drawing.**

Step 3: **Make a plan.**

Step 4: **Solve and answer the question.**

Directions Use a drawing to solve each problem. Explain the steps you took. ✎

1. Outside the Kelly Museum, there are 2 gardens: 1 circular and the other square. The distance between the centers of the gardens is 78 feet. The circular garden has a diameter of 20 feet. Each side of the square garden is 54.6 feet. How many feet apart are the edges of the 2 gardens?

   _____

2. Portia is enlarging a triangular garden. The original garden has a base of 15 feet and a height of 12 feet. Portia increases the height by 4 feet and the base by 8 feet. How much larger is the new garden?

   _____

Go on to the next page.

Using Drawings and Formulas, page 2

Drawing a picture is a helpful way to solve problems.
Helpful formulas: Area of rectangle = length × width
Area of square = side × side
Area of triangle = $\frac{1}{2}$ base × height
Area of semicircle = $\frac{1}{2}$ π × radius × radius
Use π = 3.14
Radius = $\frac{1}{2}$ diameter

Directions Use a drawing to solve each problem. Explain the steps you took. ✎

3. Teddy is painting a mural of an ice cream cone. The cone is a triangle that has a height of 24 feet and a base of 12 feet. The ice cream is a semicircle that sits on the base of the triangle. What is the approximate total area of the mural?

_____

4. Lars is walking around a path in the park. He walks 50 feet north, 40 feet east, 20 feet south, and 60 feet west. How long is the path around the park?

_____

5. The centers of 3 adjacent circles lie along the same straight line. The circles have radii of 12 inches, 16 inches, and 25 inches, in that order. How far is the center of the first circle from the center of the last?

_____

Name _____ Date _____

Estimating Area

> Sometimes an estimate is used in measuring area.
> Helpful formula: Area of a rectangle = length × width

Janice is planning to wallpaper her bedroom, her study, and her living room. How many rolls of wallpaper should she buy?

Directions → **There are many steps to this problem. Answer each question to get one step closer to the answer.** ✎

1. One wall in Janice's bedroom is 2.89 meters wide. Round this number to its nearest whole number.

   _____

2. The same wall is 3.75 meters high. Round this number to its nearest whole number.

   _____

3. Estimate the area in square meters of 1 wall in Janice's bedroom.

   _____

4. If all 4 bedroom walls have the same area, estimate the total area of the walls.

   _____

5. Each wall in Janice's study is 2.44 meters wide and 3.05 meters high. Estimate the area of the 4 walls in the study.

   _____

Go on to the next page.

Estimating Area, page 2

Sometimes an estimate is used in measuring area.
Helpful formula: Area of a rectangle = length × width

6. Two walls of Janice's living room are 3.84 meters wide and
the other two are 4.62 meters wide. If her living room is as
high as her bedroom, estimate the area of her living room walls.

_____

7. One roll of wallpaper is 0.75 meters wide and 8 meters long. This is almost
enough wallpaper to cover 1 wall in which room?

_____

8. How many rolls of wallpaper does Janice need to cover the walls in the
bedroom? in the living room?

_____

9. Explain why an estimate is an acceptable way to figure the number of rolls of
wallpaper needed. What problem might occur if the estimates are too low?

_____

_____

_____

Name _____ Date _____

Unit 4 Assessment

Directions Solve each problem. Explain your answer. ✏

1. Mrs. Smith is making a fruit salad. She bought 5 pounds of
    apples for $0.89 per pound, a pound of strawberries for $2.29
    per pound, a pineapple at $3.50, and 3 pounds of bananas at $0.39
    per pound. How much did she spend?

    _____

    _____

    _____

2. Your dog eats 14 ounces of food each day. How many pounds of food will he
    eat in a week?

    _____

    _____

    _____

3. A recipe for fruit punch calls for 3 quarts of apple juice, 2 cups of orange juice,
    1 cup of pineapple juice, and 16 ounces of sparkling water. How many 8-ounce
    glasses will it fill?

    _____

    _____

    _____

Directions Write which operation is required to solve the problem. Then,
solve. Explain why you chose the operation. ✏

4. Ashley is having a party. A snack bowl contains 96 ounces of trail mix. If there
    are 12 people at the party, how many ounces of trail mix will each person get?

    Operation _____

    Answer: _____

    _____

    _____

Choosing the Operation

Some problems do not tell you to add, subtract, multiply, or divide, or how to combine these operations. You must read the problem carefully to choose the operation.

Here's How!

Step 1: Read the problem.

Twelve people ate birthday cake. Each piece of cake weighed 0.22 pounds. One half of the cake is left. How much did the whole cake weigh?

Step 2: Choose an operation.

To find how much the whole cake weighed, multiply.

Step 3: Solve the problem.

12 pieces × 0.22 pounds per piece = 2.64 pounds
2.64 pounds × 2 halves = 5.28 pounds
The whole cake weighed 5.28 pounds.

Directions Write which operation is required to solve each problem. Then, solve each problem. Explain why you chose each operation.

1. Guests at the party drank chocolate milk. Each pitcher held 38 ounces. If 12 guests each drank 9.5 ounces of milk, how many pitchers of milk did they drink?

   Operations _____ Answer: _____

   _____

2. One snack at the party was a bag of corn chips that weighed 16.5 ounces. If 15 guests each ate an equal amount of the whole bag, how many ounces of chips did each guest eat?

   Operations _____ Answer: _____

   _____

3. Friends ate the remaining 2.64 pounds of cake. It was cut into equal pieces that weighed 0.33 pounds each. How many pieces were cut?

   Operations _____ Answer: _____

   _____

Capacity Operations

In order to add or subtract customary units of capacity, you must know how to convert them first.

Some helpful conversions: 1 cup = 8 fluid ounces
1 pint = 2 cups
1 quart = 2 pints
1 gallon = 4 quarts

Here's How!

Example 1:

$$
\begin{array}{r}
3\ \text{qt}\ 2\ \text{pt} \\
+\ 2\ \text{qt}\ 1\ \text{pt} \\
\hline
5\ \text{qt}\ 3\ \text{pt}
\end{array}
$$

Step 1: Add.

Step 2:
   Rename.   3 pt = 1 qt 1 pt

Step 3:
   Add.   5 qt + 1 qt = 6 qt

Step 4:
   Rename.   6 qt = 1 gal 2 qt

Answer: 1 gal 2 qt 1 pt

Example 2:

$$
\begin{array}{r}
3\ \text{gal}\ 1\ \text{qt} \\
-\ 2\ \text{gal}\ 2\ \text{qt} \\
\hline
\end{array}
$$

Step 1:
   Since 1 qt < 2 qt, **rename** 3 gal 1 qt.

3 gal   1 qt

2 gal + 4 qt + 1 qt

2 gal 5 qt

Step 2:
   Subtract the quarts.   2 gal 5 qt
   Subtract the gallons.   − 2 gal 2 qt
                                   3 qt

Answer: 3 qt

Directions Add or subtract.

1.   2 gal 3 qt
     + 5 gal 2 qt

2.   4 gal 1 qt
     − 1 gal 3 qt

3.   7 gal 2 qt
     − 5 gal 3 qt

4.   4 qt 1 pt
     + 1 qt 1 pt

5.   2 gal 1 qt
     + 2 gal 3 qt

6.   2 gal 1 c
     + 1 gal 1 c

Capacity Word Problems

When working in the kitchen, you must know how to correctly convert capacity.

Some helpful conversions:

    1 cup = 8 fluid ounces   1 quart = 2 pints

    1 pint = 2 cups   1 gallon = 4 quarts

Here's How!

Step 1: Read the problem.

A recipe for soup uses 1 quart of beef broth, 1 pint of tomato juice, and 1 cup of milk. How many 1-cup servings will it make?

Step 2: Convert each measurement to cups.

1 quart = 2 pints = 4 cups beef broth

1 pint = 2 cups tomato juice

1 cup milk

Step 3: Add all cups together.

4 cups + 2 cups + 1 cup = 7 cups. It will make 7 servings.

Directions Solve each problem. Show your conversions on your own paper.

1. A recipe for fruit punch uses 5 quarts of pineapple juice, 3 cups of cranberry juice, and 2 pints of orange juice. How many 8-fl oz glasses will it fill?

_____

2. A gravy recipe calls for 3 pints of milk, 2 cups of chicken broth, and 4 cups of vegetable stock. How many $\frac{1}{2}$-cup servings will this recipe make?

_____

3. A recipe for perfume calls for 10 fl oz of rose oil, 12 fl oz of lavender oil, and 26 fl oz of chamomile extract. How many 6-fl oz bottles will this recipe fill?

_____

4. A baby-sitter wants to give each of the 3 children the same amount of juice. The juice comes in a 45-fl oz container. How much can each child get?

Organizing Data

Problems sometimes have many pieces of data. It is helpful to organize the data in order to simplify the problem.

Here's How!

Step 1: Read the problem.

The seniors at Van Buren High School are washing cars to raise money for a trip. It takes them 15 minutes for a car, 20 minutes for a van, and 30 minutes for a pick-up truck. How much time will it take the seniors to wash 9 cars, 5 vans, and 8 trucks?

Step 2: Organize your data.

9 cars at 15 minutes each

5 vans at 20 minutes each

8 trucks at 30 minutes each

Step 3: Solve.

Multiply the number of vehicles by the amount of time needed to wash them.

$15 \times 9 = 135$ $20 \times 5 = 100$ $30 \times 8 = 240$

Add the times together to get a total amount of time.

$135 + 100 + 240 = 475$ minutes or 7 hours and 55 minutes

It will take the seniors 7 hours and 55 minutes to wash all the vehicles.

Directions Solve each problem. Explain your answer.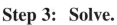

1. Mrs. Gross bought a 20-lb turkey at $0.55 per pound. She also bought 2 bags of stuffing at $1.79 each and 8 potatoes at $0.35 each. How much change did she get from $20.00?

_____

_____

2. The juniors sold fruit at $1.29 per pound. Mr. Duffy bought 10 pounds, Ms. Levy bought 8 pounds, and Ms. Lopez bought 5 pounds. How much money did the juniors earn?

_____

_____

Go on to the next page.

Organizing Data, page 2

Problems sometimes have many pieces of data. It is helpful to organize the data in order to simplify the problem.

Here's How!

   Step 1: Read the problem.

   Step 2: Organize your data.

   Step 3: Solve.

Directions   Solve each problem. Explain your answer. ✎

3. When Marcy calls her Aunt Dora, the call costs $1.25 for the first minute and $0.45 for each additional minute. How much would a 10-minute call cost?

_____

_____

_____

4. Sally had T-shirts for sale at a fair. On Friday, she sold each shirt for $8 and made $56. On Saturday, she reduced the price to $6 each and sold 4 times as many. How much money did she earn altogether?

_____

_____

_____

5. Chris is buying video games with his birthday money. He buys 3 games that cost $10 each, 2 games that cost $20 each, and a controller that costs $15. How much money will he spend altogether?

_____

_____

_____

Name _____   Date _____

Flight Schedules

The airline schedules below show flight times from Mexico City to Birmingham, Alabama, and from Birmingham to Kansas City, Missouri.

From Mexico City to Birmingham		
Depart	Arrive	Flight No.
7:50 A.M.	3:40 P.M.	926
12:48 P.M.	8:35 P.M.	904

From Birmingham to Kansas City		
Depart	Arrive	Flight No.
7:05 A.M.	10:45 A.M.	104
9:55 A.M.	1:11 P.M.	148
2:50 P.M.	6:47 P.M.	284
4:37 P.M.	8:17 P.M.	784

Here's How!

Step 1: Read the problem.

What is the flying time from Mexico City to Birmingham on Flight 926?
Hint: When subtracting times, it is helpful to change the times to 24-hour time. Add 12 hours to the hours place for afternoon times.

Step 2: Subtract the departure time from the arrival time. Solve.

$$
\begin{array}{r}
3 \text{ h } 40 \text{ min (P.M.)} \\
- \ 7 \text{ h } 50 \text{ min (A.M.)}
\end{array}
\longrightarrow
\begin{array}{r}
\overset{14}{\cancel{15}} \text{ h } \overset{100}{\cancel{40}} \text{ min} \\
- \ 7 \text{ h } 50 \text{ min}
\end{array}
\longrightarrow
\begin{array}{r}
14 \text{ h } 100 \text{ min} \\
- \ 7 \text{ h } \ 50 \text{ min} \\
\hline
7 \text{ h } \ 50 \text{ min}
\end{array}
$$

Directions   Solve each problem. ✎

1. What is the flying time from Birmingham to Kansas City on Flight 104?

_____

_____

2. Find the flying time from Birmingham to Kansas City on Flight 284.

_____

_____

3. How much later does Flight 784 leave Birmingham than Flight 284?

_____

_____

Go on to the next page.

Name _____ Date _____

Flight Schedules, page 2

From Mexico City to Birmingham		
Depart	Arrive	Flight No.
7:50 A.M.	3:40 P.M.	926
12:48 P.M.	8:35 P.M.	904

From Birmingham to Kansas City		
Depart	Arrive	Flight No.
7:05 A.M.	10:45 A.M.	104
9:55 A.M.	1:11 P.M.	148
2:50 P.M.	6:47 P.M.	284
4:37 P.M.	8:17 P.M.	784

Directions Solve each problem. ✐

4. How long does it take to fly from Birmingham to Kansas City on Flight 148?

_____

_____

_____

_____

5. Sandra flies from Mexico City to Birmingham on Flight 926. The next day she flies from Birmingham to Kansas City on Flight 148. What is her total flying time from Mexico City to Kansas City?

_____

_____

_____

_____

6. The Morans fly to Birmingham from Mexico City on Flight 904. Later they fly from Birmingham to Kansas City on Flight 784. What is their total flying time?

_____

_____

_____

_____

Name _____   Date _____

Planning a Budget

You are the set designer for a play. Below is a list of prices of the supplies that you need.

paint $9.76 per gallon   wood $1.68 per foot
fabric $4.15 per yard   nails $1.89 per box
canvas $30.45 per roll   rope $2.30 per yard
brushes $5.20 each   masking tape $0.90 per roll

Directions Solve each problem. Explain your answer.

1. A lumberyard has offered to supply $200.00 worth of free merchandise. Your design calls for 100 feet of wood and 12 boxes of nails. Could you get 5 more feet of wood without having to pay the lumberyard any money?

   _____

   _____

   _____

2. You need 76 yards of fabric and 9 rolls of canvas. The store gives a discount on any order that totals more than $600.00. How many more rolls of canvas would you have to buy to qualify for a discount?

   _____

   _____

   _____

3. You have brought $300.00 to the hardware store. You want to buy 21 gallons of paint, 6 brushes, and 12 rolls of masking tape. Do you have enough money to buy 30 yards of rope as well?

   _____

   _____

   _____

Go on to the next page.

Planning a Budget, page 2

You are the set designer for a play. Below is a list of prices of the supplies that you need.

paint	$9.76 per gallon	wood	$1.68 per foot
fabric	$4.15 per yard	nails	$1.89 per box
canvas	$30.45 per roll	rope	$2.30 per yard
brushes	$5.20 each	masking tape	$0.90 per roll

Directions Solve each problem. Explain your answer. ✎

4. You wrote checks for 5 yards of fabric to Cloth Cutters, Inc., and for 10 yards of rope to Ward Hardware. You forgot to fill in the names. Which store would have received the check for the larger amount?

_____

_____

5. A friend has asked you to order 2 gallons of paint, 1 paint brush, and 54 feet of wood. He wants to spend about the same amount on canvas as he spends on the above items. How many rolls of canvas should you order for him?

_____

_____

6. You need 4 new curtains. Each will use 12 yards of fabric, and each will need 10 yards of rope. Your budget is $275.00. Will this be enough?

_____

_____

7. To make a sign for the play, you need 6 yards of rope, 1 roll of canvas, 2 gallons of paint, and 2 brushes. How much money will you have left if you have $100.00 to spend?

_____

_____

Weight Word Problems

To measure weight, use ounces, pounds, and tons. To change one unit of measure to another, multiply or divide.

Helpful conversions: 1 pound (lb) = 16 ounces (oz)
                                 1 ton (T) = 2,000 pounds (lb)

To change pounds to a smaller unit such as ounces, multiply.
        1 lb = 16 oz
        3 lb × 16 oz = 48 oz
        3 lb = 48 oz

To change pounds to a larger unit such as tons, divide.
        2,000 lb = 1 T
        4,000 lb ÷ 2,000 lb = 2 T
        4,000 lb = 2 T

Here's How!

Step 1: Read the problem.
Your dog weighs 14 pounds. How many ounces does your dog weigh?

Step 2: Identify what the problem asks.
The problem asks how many ounces are in 14 pounds.

Step 3: Solve the problem.
Since ounces is a smaller unit, multiply 14 pounds by 16 ounces.
14 lb × 16 oz = 224 oz
Your dog weighs 224 ounces.

Directions Solve each problem. Explain how you changed units. ✏️

1. You buy dog food that weighs 160 ounces. How many pounds are in the bag?

_____

_____

Go on to the next page.

Weight Word Problems, page 2

To measure weight, use ounces, pounds, and tons. To change one unit of measure to another, multiply or divide.

Helpful conversions: 1 pound (lb) = 16 ounces (oz)
1 ton (T) = 2,000 pounds (lb)

Directions Solve each problem. Explain how you changed units. ✏️

2. You catch a fish that weighs 32 ounces. How many pounds is this?

_____

_____

3. The neighbor's cat eats 8 ounces of food each day. How many pounds of food will the cat eat in 8 days?

_____

_____

4. The zoo elephant weighs about 3 tons. How many pounds is this?

_____

_____

5. The neighbor's cat weighs 10 pounds. How many ounces is this?

_____

_____

6. The concrete used for a driveway weighs 10,000 pounds. How many tons is this?

_____

_____

Name _____ Date _____

Unit 5 Assessment

Directions Solve each problem. Explain your answer.

1. Matthew has 5 shirts and 4 ties that match. How many combinations of one shirt and one tie can he make?

   _____

   _____

   _____

2. Very Cold Cola is coming out with a new cola. The company surveyed 100 people outside the Very Cold Cola building. Of those surveyed, 85 people loved it. Was this a good sample? Explain.

   _____

   _____

Directions Use the graph for problems 3 and 4.

3. Between which two months was there the largest decrease in attendance percentage?

   _____

4. In which month was the attendance the lowest?

   _____

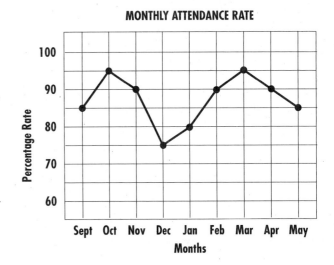

MONTHLY ATTENDANCE RATE

Line Graphs

Line graphs are often used to show how data changes over time. This graph shows the school attendance rate by month.

Here's How!

Step 1: Put your finger on the month on the bottom of the graph.

Step 2: Move your finger up to the dot for that month.

Step 3: Move your finger to the left to find the percentage rate.

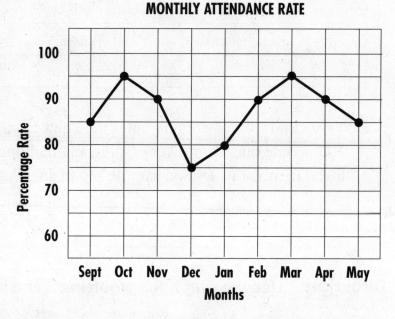

MONTHLY ATTENDANCE RATE

Directions Use the graph to answer the questions. ✎

1. How much greater was attendance in September than in December?

   _____

2. Attendance was how much greater in the highest month than in the lowest month?

   _____

3. In what 3 months was the attendance rate the same?

   _____

4. Was the attendance rate higher in September–December or February–May?

   _____

Double Line Graphs

Double line graphs are often used to compare monthly information for
different years. This double line graph compares the monthly attendance
for 2 years.

Here's How!

Step 1: **Read the solid line
for 1999 data.**

Step 2: **Read the dotted line
for 2000 data.**

Step 3: **Compare the data to
answer the questions.**

———— 1999
- - - - 2000

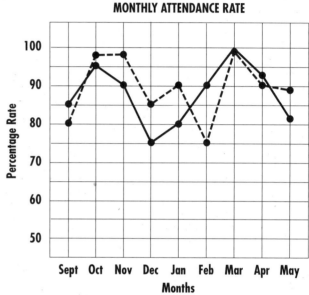

MONTHLY ATTENDANCE RATE

Directions Use the graph to answer the questions. ✎

1. In which year was the overall attendance rate higher?

_____

2. In which months was the 2000 rate lower than the 1999 rate?

_____

3. When was the attendance rate the lowest for each year?

_____

4. In which month was the attendance rate the same for both years?

_____

Circle Graphs

A circle graph is also called a pie graph because it looks like a pie. Each slice of the pie is a piece of information included in the whole pie. The size of each piece shows its relationship to the whole pie and to other pieces of the pie.

Here's How!

Step 1: Study the Monthly Budget graph.

Step 2: Compare pieces of the pie.

Step 3: Compare a piece of the pie to the whole pie.

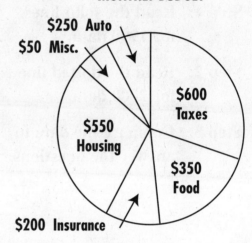

MONTHLY BUDGET

$250 Auto
$50 Misc.
$600 Taxes
$550 Housing
$350 Food
$200 Insurance

Directions Use the circle graph to answer the questions. ✐

1. What is the total budget in this graph?

   _____

2. Which two expenses equal almost half of the total budget?

   _____

3. Which expenses are similar in amount?

   _____

4. What is the difference between the greatest expense and the least expense?

   _____

Name _____ Date _____

More Circle Graphs

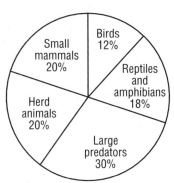

Animals at the Colorado Canyon Zoo (700 animals in all)

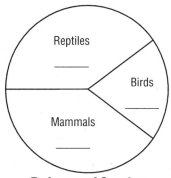

Endangered Species (140 animals in all)

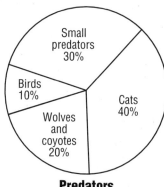

Predators (210 animals in all)

Directions → Use the circle graphs to answer each question. Explain your answer. ✐

1. How many reptiles and amphibians are there at the zoo?

   _____

   _____

2. Of the herd animals at the zoo, 36 are giraffes. How many of the herd animals are not giraffes?

   _____

   _____

3. How many more herd animals are there than reptiles and amphibians?

   _____

   _____

4. There are 28 birds in the endangered species section at the zoo. What percent is this? Write the percent on the graph.

   _____

   _____

Go on to the next page.

More Circle Graphs, page 2

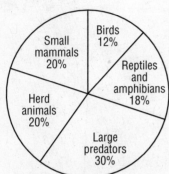

Animals at the Colorado Canyon Zoo (700 animals in all)

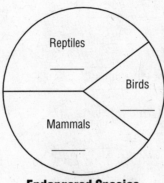

Endangered Species (140 animals in all)

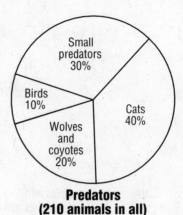

Predators (210 animals in all)

Directions Use the circle graphs to answer each question. Explain your answer.

5. Of the endangered species, 40% are mammals. Of the predators, 30% are small predators. Which group has more animals, endangered mammals or small predators? How many more?

_____

6. What is the percent of endangered reptiles at the zoo? Write the percent on the graph.

_____

7. How many endangered reptiles are there?

_____

8. How many wolves and coyotes are among the predators at the zoo? Is this more or less than the number of endangered birds?

_____

9. Which bird group is larger, predators or endangered species? How much larger?

_____

Combinations

> Combinations are made by finding all the different ways to group a set of items. One way to find all the different combinations of a set of items is to make a tree diagram. Another way is to multiply the number of choices together.

Here's How!

Mark has 4 sweaters and 3 shirts that can be worn in any combination. How many shirt-sweater combinations can he make?

Method 1		**Method 2**

Method 1

red sweater
- red shirt — 1
- blue shirt — 2
- green shirt — 3

blue sweater
- red shirt — 4
- blue shirt — 5
- green shirt — 6

white sweater
- red shirt — 7
- blue shirt — 8
- green shirt — 9

green sweater
- red shirt — 10
- blue shirt — 11
- green shirt — 12

Method 2

sweaters   shirts

$4 \times 3 = 12$

Directions Choose a method to answer each question. Explain the steps you took to get your answers. You may use your own paper.

1. Dustin has 5 kinds of crackers and 3 kinds of cheese. How many cracker-cheese combinations can he make?

_____

_____

Go on to the next page.

Combinations, page 2

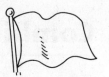

Combinations are made by finding all the different ways to group a set of items. One way to find all the different combinations of a set of items is to make a tree diagram. Another way is to multiply the number of choices together.

Directions **Choose a method to answer each question. Explain the steps you took to get your answers. You may use your own paper.** ✏️

2. A sailor can send messages by displaying flags of different colors. He has a flagpole that has room for 3 flags. He has 4 flags that can be used for the top position, 3 flags that can be used for the middle position, and 2 flags that can be used for the bottom position. How many combinations of flag positions can he use?

_____

_____

_____

3. Good Eats Diner is serving a choice of 4 meats, 4 vegetables, and 2 desserts. How many meal combinations of 1 meat, 1 vegetable, and 1 dessert can be made?

_____

_____

_____

4. Students at Greenwood Intermediate School are holding student council elections. Three people are running for president, 3 for vice-president, 2 for treasurer, and 2 for secretary. How many possible combinations of student council officers are there?

_____

_____

_____

Name _____ Date _____

Sampling

When you analyze a group of things, it is sometimes more practical to look at a small section of the group. This is called a sampling. In order for the data to be accurate, the sample group should be as similar as possible to the group as a whole. The larger the sample group, the more accurate the results of the survey will be. Also, in order to obtain a good sample, a variety of people should be surveyed.

Souper Soup Company thinks that its New Carrot Soup will be very popular. The company took a survey of its employees to see what people thought about the soup.

Directions Answer the questions. Explain each answer. ✏️

1. They asked 4 of the 5,000 employees. All 4 loved the soup. The company said that everyone loved the soup. Was this a good sample? Explain.

    _____

    _____

2. Next, the company surveyed 1,000 of its employees, and 900 said they loved the New Carrot Soup. The company decided that 9 of every 10 people would love the soup. Was this more accurate than the other sample? Explain.

    _____

    _____

3. Souper Soup surveyed 1,000 employees of the Simmering Soup Company, a rival soup maker. New Carrot Soup was not a hit. Only 25 people liked the soup; 975 did not like it. What was wrong with the sample?

    _____

    _____

4. What would be a good sample group for Souper Soup to use to test New Carrot Soup?

    _____

    _____

Name _____ Date _____

More Sampling

Albert has just taken a job as an egg spot-checker. Albert must check 2 cartons out of every 100 that the company ships to stores. He checks for cracked eggs, empty egg pockets, and other problems.

Directions Answer the questions below. Explain each answer. ✑

1. On Monday, Albert checks 2 cartons from the top of every crate of 100. Is his sampling accurate?

   _____

   _____

2. On Tuesday, Albert checks 200 of the 10,000 cartons that are shipped. He finds 10 with broken eggs in them. He reports that only 10 imperfect cartons were shipped. Is his reasoning correct?

   _____

   _____

3. On Wednesday, Albert accidentally drops a crate of eggs. When he opens the first carton from the crate, half of the eggs are broken. So are half of the eggs in the second carton. He reports that half of the eggs shipped that day were broken. Is this accurate?

   _____

   _____

4. On Friday, Albert checks 100 cartons out of 5,000. He finds only one broken egg. He reports 50 broken eggs in 5,000 cartons. Is his report accurate?

   _____

   _____

Multistep Math Problems with Written Explanations, Grade 5

Answer Key

Note: Answer explanations will vary. Accept all reasonable explanations. Possible answers are given.

pp. 3–4
1. Yes, her total will be $98.94.
2. 61 boxes; $x + 25 = 86$
3. Yes, he has enough money. He has $1.80.
4. 2 quarters, 3 dimes, 2 nickels, 4 pennies
5. 52.74 square yards; The area of the pond is 28.26 square yards. The area of the square is 81 square yards. $81 - 28.26 = 52.74$
6a. 33 square meters; Multiply height by width for each wall. Since there are 2 of each size wall, multiply each answer by 2. Add the 2 numbers together to get all 4 walls.
6b. 6 rolls; Multiply the height and width to get the total area of 1 roll of wallpaper. $0.75 \times 8 = 6$ square meters. Divide the total area of the walls by the total area of one roll of wallpaper. $33 \div 6 = 5$ R 3. The remainder tells us to buy another roll.
7. 1 pound; 112 ounces $\div 7$ days = 16 ounces per day. Since 16 ounces equal 1 pound, the dog eats 1 pound of dog food per day.

p. 5
1. $77.94; $12.99 \times 2 = 25.98 $3.99 \times 3 = 11.97 Add these 2 numbers to $39.99 to find the total amount she will spend.
2. 8 boxes; 38 books $\div 5$ books per box = 7 R3. The remainder tells us we need an additional box.
3. 12 miles; Add the distances for Monday and Wednesday. $11 + 12 = 23$ miles; Subtract the sum from the total distance to get the distance on Friday. $35 - 23 = 12$
4. Aaron should choose the second plan. It will save him $10.00. $3.00 \times 20 = $60.00; $10.00 \times 5 = 50.00

p. 6
1. 67; Add 25, 30, and 20. Subtract 8 from the result.
2. $3; Multiply 55 by $4 and 55 by $5. Then, add the two totals together. Next, subtract that result from $600. Finally, divide that amount by 35.
3. 82; Add 19, 16, and 28. Then, add 55, 35, and 55. Finally, subtract the first total from the second.

pp. 7–8
1. 5; 6 people were in the original group. Subtract 1 when Richard left and add 2 when the friends came. There are now 7 people. Half of the original group is 3, so 4 more people left. $4 + 1 = 5$ people total.

2. 12; 45 noodles $\times 2$ meters each = 90 meters. 90 meters $\div 8$ meters per spool = 11 R2. The remainder tells us that Mark needs one more spool.
3. $12.50; Add the prizes for the scariest costume and the funniest costume. Divide by 2 people to find out how much each won.
4. $8; $8.50 \times 2 = $17; $25 - $17 = 8

pp. 9–10
1. Step 1: 4 days $\times 10$ miles per day = 40 miles; Step 2: 60 miles – 40 miles = 20 miles; Answer: 20 miles
2. Step 1: 20 miles + 42 miles = 62 miles; Step 2: 62 miles – 50 miles = 12; Answer: 12 miles
3. 20 miles; 4 days $\times 10$ miles per day = 40 miles; 60 miles – 40 miles = 20 miles saved
4. Tates; 4 days $\times 7$ miles per day = 28 miles; 28 miles + 14 miles = 42 miles; The Tates' route is 42 miles through the mountains.
5. Yes; 10 miles per day $\times 7$ days = 70 miles; The Tates' route is 50 miles to the river plus 20 miles to the mountains.
6. 6 miles; Add $60 + 30 + 28 = 118$ miles to get the Chans' route. Add $50 + 20 + 42 = 112$ miles to get the Tates' route. Subtract to find how much longer the Chans' route is.

pp. 11–12
1. $51.30; $895.00 + $53.70 = 948.70 for the computer and tax; Subtract the result from his savings to find out how much he has left. $1,000.00 - $948.70 = 51.30
2. $7.75; Add the costs of the program and mouse. Subtract the result from $50.00 to see how much she has left.
3. $1.11; Add the costs of the paper, magazine, and disks. Subtract the result from $20.00 to find her change.
4. 52 disks; Add Elena's disks to Ethan's disks. Subtract the result from 100 to find how many are left.
5. $42.33; Add her savings to her earnings. Subtract from the cost of the printer to see how much more she needs.
6. Yes; Multiply $19.95 by 2 for the first two games and add $29.95 to the result. $69.85 is less than $100.00.

pp. 13–14
1. 8, 2; 45 $\div 2 = 8$ R 2. The remainder tells that there will be 2 tanks left over.
2. 5; 16 hours $\div 3 = 5$ R 1. The remainder tells that there will be 1 hour left over.
3. 9; 68 $\div 8 = 8$ R 4. The remainder tells that there will be 4 pounds of rocks left over. There must be another trip to bring the last 4 pounds.

pp. 13–14 (cont.)

4. 8, 3; $35 \div 4 = 8$ R 3. Since it takes 4 packs to go to and from the site, there can only be 8 trips. There will be 3 fuel packs left over.

5. 8, 4; $60 \div 7 = 8$ R 4. The 4 tells there will be 4 pictures left over and that there can be 8 groups of 7 pictures.

6. 5 experiments per person; $22 \div 4 = 5$ R 2; The remainder tells that there will be 2 experiments left. 3 per day; $22 \div 7 = 3$ R 1. The remainder tells that there will be 1 experiment left to do. 2 people; 4 experiments on the last day

p. 15

1. $0.35; Subtract the amount Daniel spent from the amount Angelo spent, then divide by 2.

2. 500 meters; Multiply the length of the track by the number of laps for each race, then subtract to find the difference.

p. 16

1. $46.13; Multiply the number of bills or coins by their values. Add the values together. Do this for both boys. Add to find the total value.

2. 1 penny, 6 nickels, 5 dimes

3. 16 caps; Let x = the number of caps Eric has.; $x - 4 = 12$ or $x = 12 + 4$; $x = 16$

4. 16 days; Let x = the number of days he walked the dogs.; $4x = 64$; $64 \div 4 = x$; $x = 16$

pp. 17–18

1. Brett has earned $15.80 and Stephen has earned $15.70. Brett has earned more money. Make a model to show how much each boy has earned. Compare the amounts.

2. $5.62; Make a model to show how much each boy has saved. Add to find the total amount of money saved.

3. Yes. Add the cost of the shoes to the cost of 1 game of bowling to find the total cost ($5.00). Next, make a model to find out how much money she has with her ($5.03). Compare the result to the total cost of bowling.

4. Jeff can buy only 1 half-gallon. Make a model to calculate how much money Jeff has. He has $2.96. Since 2 half-gallons cost $3.98, he can buy only 1.

5. Nabil had $0.65 in change. He could buy 1 or 2 cookies for dessert.

pp. 19–20

1. 8 nickels, 4 dimes, 2 quarters
2. 39 dimes, 60 pennies, 1 half-dollar
3. $0.36; 3 nickels, 2 dimes, and 1 penny
4. 2 quarters, 20 dimes, 10 nickels
5. 4 quarters, 4 dimes, 12 nickels
6. 8 dimes, 32 nickels, 10 pennies

p. 21–22

1. 22 red beads; Let r = the number of red beads.; $r + 8 = 30$; $r + 8 - 8 = 30 - 8$

2. 78 pounds; Let T = Tom's weight.; $T + 9 = 87$; $T + 9 - 9 = 87 - 9$

3. 10 hours; Let S = the number of hours she worked in September.; $S + 5 = 15$; $S + 5 - 5 = 15 - 5$

4. 5 chapters; Let x = the number of chapters he read on Saturday.; $x + 3 = 8$; $x + 3 - 3 = 8 - 3$

5. 90 points; Let p = the number of points received if on time; $p - 4 = 86$; $p - 4 + 4 = 86 + 4$

pp. 23–24

1. $8.00; Let x = the cost of oil.; $x + 18 = 26$; $x = 26 - 18$

2. $208.00; Let t = the cost of the tackle altogether.; $t = 4 \times 52$

3. 75 feet; Let r = the length of each new rope.; $r = 225 \div 3$

4. 9 miles; Let m = the number of miles per day.; $m = 81 \div 9$

5. 56; Let t = the number of total miles.; $t = 11.2 \times 5$

pp. 25–26

1. 8 minutes 55.3 seconds; $x = 2$ min 23.3 sec + 6 min 32 sec

2. 2.51 million copies; $x = 1.35 + 0.42 + 0.74$

3. 46,655 fans; $x = 47,320 - 2,420 + 1,755$

p. 27

1. 60 in.; 14 in. $\times$ 2 = 28 in.; 16 in. $\times$ 2 = 32 in.; 28 + 32 = 60 in.

2. circumference = 31.4 in.; area = 78.50 square in.

3. 586 square feet; Find the area of the circle and the square. Subtract the circle's area from the square's area. (900 – 314)

4. 200 square ft; Multiply the length by the width.; $20 \times 10 = 200$

p. 28

1. 3,108 feet; Add the lengths of the sides, then multiply by 4.

2. 5 feet by 8 feet; 26 feet; Find the largest board for each side.; Add the lengths of the sides.

3. 27 sections; Find how many sections for each side, then multiply by 3 sides.

pp. 29–30

1. Circumference = 56.52 inches; Area = 254.34 square inches
2. Circumference = 50.24 inches; Area = 200.96 square inches
3. Area = 113.04 square inches
4. Area = 78.5 square inches
5. 9.8 square inches
6. 30 inches

pp. 31–32

1. 123.84 square cm
2. 123.84 square cm
3. 123.84 square cm
4. 16 circles
5. 123.84 square cm
6. 36 circles
7. Answers will vary.

pp. 33–34

1. 452.16 square feet
2. 350 square inches
3. 226.08 square feet
4. 180 feet taller
5. greater by 2,200 square feet

pp. 35–36

1. 40.7 feet; Take half of the diameter to find the radius of the circle. Subtract that amount from the distance between the centers. Take half of the length of the side and subtract it from the previous result.
2. 94 square feet; Find the area of the original garden. Add the new amounts to the base and height and find the area of the new garden. Subtract the area of the old from the area of the new.
3. 200.52 square feet; Find the area of the triangle and semicircle. Add the two areas together.
4. 170 feet; Add all the distances together.
5. 69 inches; Add the radius of the first circle, the diameter of the second circle, and the radius of the third circle.

pp. 37–38

1. 3 meters
2. 4 meters
3. about 12 square meters
4. about 48 square meters
5. about 24 square meters
6. about 72 square meters
7. the study
8. 8 rolls; 12 rolls
9. You cannot buy a partial roll of wallpaper.; You might not have enough wallpaper.

p. 39

1. $11.41; Multiply 5 pounds apples by $0.89 and 3 pounds bananas by $0.39. Add these amounts to $2.29 for strawberries and $3.50 for pineapple.
2. 6 pounds 2 ounces; Multiply 14 ounces by 7 days. Divide the result by 16 ounces per pound. The remainder is 2 ounces.
3. 17 glasses; Convert each capacity to ounces. Add the result. Divide that result by 8 ounces.
4. division; 8 ounces; A large quantity needs to be split between a few people.

p. 40

1. multiplication and division; 3 pitchers; Multiply the number of guests times 9.5 ounces. Divide the result by the number of ounces in one pitcher.
2. division; 1.1 ounces each; Divide the amount of corn chips by the number of guests.
3. division; 8 pieces; Divide the cake into 0.33 pound pieces.

p. 41

1. 8 gal 1 qt
2. 2 gal 2 qt
3. 1 gal 3 qt
4. 1 gal 2 qt
5. 5 gal
6. 3 gal 1 pt

p. 42

1. 27 glasses
2. 24 servings
3. 8 bottles
4. 15 fl oz

pp. 43–44

1. $2.62; Multiply the number of pounds in the turkey by the price per pound. Multiply the cost of stuffing per bag by the number of bags. Multiply the number of potatoes by the cost per potato. Add these numbers together. Subtract the result from $20.00.
2. $29.67; Multiply the number of pounds that each person bought by $1.29 per pound. Add the results together.
3. $5.30; $0.45 × 9 for the last nine minutes. Add $1.25 for the first minute.
4. $224; Divide $56 by $8 to find the number of shirts sold on Friday. Multiply that number by four. Then, multiply the result by $6. Add that amount to $56.
5. $85; Multiply the number of games he buys by their appropriate costs. Add the results together. Add $15 for the controller.

pp. 45–46

1. 3 h 40 min
2. 3 h 57 min
3. 1 h 47 min
4. 3 h 16 min
5. 11 h 6 min
6. 11 h 27 min

p. 47–48

1. Yes; The cost of 100 feet of wood and 12 boxes of nails comes to $190.68. The cost of an additional 5 feet of wood would bring the total to $199.08.
2. 1 roll; The cost of 76 yards of fabric and 9 rolls of canvas is $589.45. The cost of an additional roll of canvas brings the total to more than $600.00.

p. 47–48 (cont.)

3. No; The cost of 21 gallons of paint, 6 brushes, and 12 rolls of masking tape comes to $246.96. The cost of 30 yards of rope brings the total to more than $300.00.
4. Ward Hardware; $5 \times \$4.15 = \$20.75$; $10 \times \$2.30 = \$23.00$
5. 4 rolls; The supplies cost $115.44. 4 rolls of canvas cost about the same amount.
6. No; 4 curtains would cost $291.20.
7. $25.83; The materials cost $74.17. Subtract that cost from $100.00 to find the money left from the original amount.

pp. 49–50

1. 10 pounds; Divide the ounces by 16 to change it to pounds.
2. 2 pounds; Divide the ounces by 16 to change it to pounds.
3. 4 pounds; Multiply the number of ounces per day by the number of days. Divide the result by 16 to change it to pounds.
4. 6,000 pounds; Multiply 3 tons by 2,000 to change it to pounds.
5. 160 ounces; Multiply 10 pounds by 16 to change it to ounces.
6. 5 tons; Divide the number of pounds by 2,000 to change it to tons.

p. 51

1. 20 combinations; Multiply the number of shirts by the number of ties.
2. No. The sample needs to be done away from the Very Cold Cola building. Many people near the building would work there.
3. Nov.–Dec.; The largest drop occurs between November and December.
4. Dec.; The attendance was the lowest in the month of December.

p. 52

1. 10%
2. 20%
3. November, February, April
4. February–May

p. 53

1. 2000
2. September, February, and April
3. December 1999 and February 2000
4. March

p. 54

1. $2,000.00
2. Taxes and Food

3. Auto and Insurance, Taxes and Housing
4. $550; Taxes – Misc.

pp. 55–56

1. 126 reptiles and amphibians; $700 \times 0.18 = 126$
2. 104 are not giraffes. $700 \times 0.20 = 140$; $140 - 36 = 104$
3. 14 more; $140 - 126 = 14$
4. 20%; $28 \div 140 = 0.20$
5. There are 7 more small predators. $140 \times 0.40 = 56$; $210 \times 0.30 = 63$; Compare the numbers.
6. 40%; Add the percentages of mammals and birds. Subtract the result from 100.
7. 56; $140 \times 0.40 = 56$
8. 42; $210 \times 0.20 = 42$; more; There are 28 endangered birds at the zoo.
9. endangered species; $210 \times 0.10 = 21$; $140 \times 0.20 = 28$. Compare these numbers.

p. 57–58

1. 15; Multiply the number of cracker choices by the number of cheese choices.
2. 24; Multiply the number of flags that will go in each position.; $4 \times 3 \times 2 = 24$
3. 32; Multiply the number of choices of meats by the number of choices of vegetables and then by the number of choices of desserts.; $4 \times 4 \times 2 = 32$
4. 36; Multiply the number of candidates for president by the number of candidates for vice president and then by the number of candidates for treasurer and then by the number of candidates for secretary. $3 \times 3 \times 2 \times 2 = 36$

p. 59

1. No. Employees of the company may feel obligated to like the soup. The size of the sample was too small.
2. Yes. It was more accurate than the previous sample. However, the people surveyed should be chosen at random.
3. The employees of a rival company would feel loyalty to their company and may not give a completely honest answer.
4. A good sample group would be 1,000 people chosen at random.

p. 60

1. No.; The eggs on the bottom are more likely to get crushed. A few cartons in all areas should be tested.
2. No.; He needs to use the sample data and use a proportion to figure approximately how many eggs would be broken.
3. No.; not all crates were dropped
4. Yes.; One egg out of 100 cartons is proportional to 50 eggs in 5,000 cartons.